Obscure Geometry

by JonS

DORRANCE
PUBLISHING CO
EST. 1920
PITTSBURGH, PENNSYLVANIA 15238

Dorrance Publishing Co
585 Alpha Drive
Suite 103
Pittsburgh, PA 15238
Visit our website at *www.dorrancebookstore.com*

ISBN: 979-8-8860-4156-9
eISBN: 979-8-8860-4812-4

Obscure Geometry

Introduction

At an early age of 3 my family pulled up stakes from our Massachusetts home and moved to the Island of Trinidad & Tobago; the birthplace of My Mother and Her Family. My Folks purchased a beach house on the North Coast of the island in a small village. The property was known as Pelican House with a great view of the ocean where stood a large rock formation some 200 feet off shore where pelicans gathered. A large Almond tree stood between our house and the ocean where one could find me on any given day husking almonds with a rock while barefoot and in shorts. We being the only bi-racial family in the village raised a few eyebrows at first but soon the villagers became warm and receptive to their new neighbors. A wonderful Hindu family lived next to us with three young Sons and a teen age daughter. We soon became close friends. They shared a great deal of their culture with us from climbing Coconut Palms to fetch the nuts to trapping land crabs with their bamboo hollowed out traps placed in the crab holes at night to trapping song birds with tree gum to show off with their home made cages. We returned to The States after a year but retained the property for future visits. We purchased a house in New Hampshire where we still live. When I was 9 my folks separated and my Dad bought a home in Northern Maine with a lovely view of Mt Katahdin and the Mountain range. I've spent countless hours soaking up the view and long walks at night enjoying the star lit nights that appear so close. Several years later at 13 I developed an interest in playing the harp and that is part of my life to this day.

All the diverse experiences has brought me to this metaphysical place that explains the genesis of Obscure Geometry. One evening while on the living room couch I focused on a large star wall ornament and at that moment a feeling came over me like none other and hexagon figures appeared before me with numbers attached as presented in the manuscript in that order, like an Epiphany.

I believe there is a greater meaning when the figures are melded together as one and I await that moment.

JonSebastian Fitton

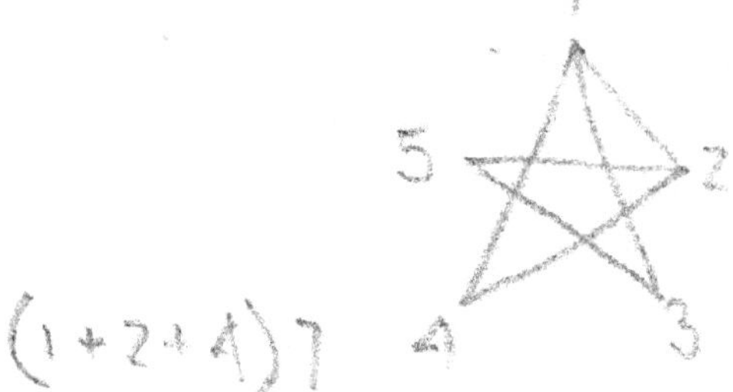

$(1 + 2 + 4) \, 7$

$(2 + 3 + 5) \, 10$

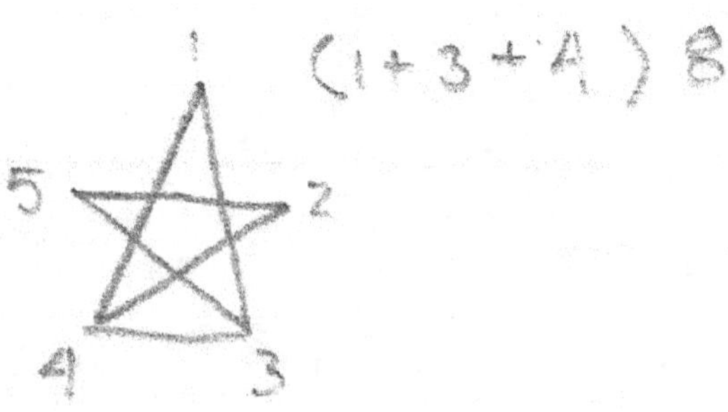

1
5   2
4   3
(1 + 3 + 4) 8

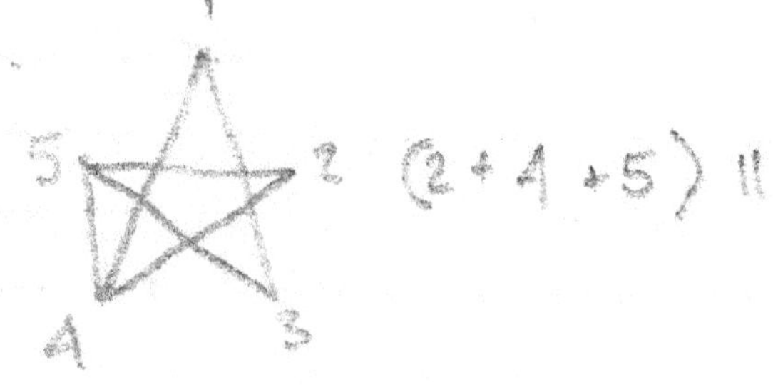

1
5   2
4   3
(2 + 4 + 5) 11

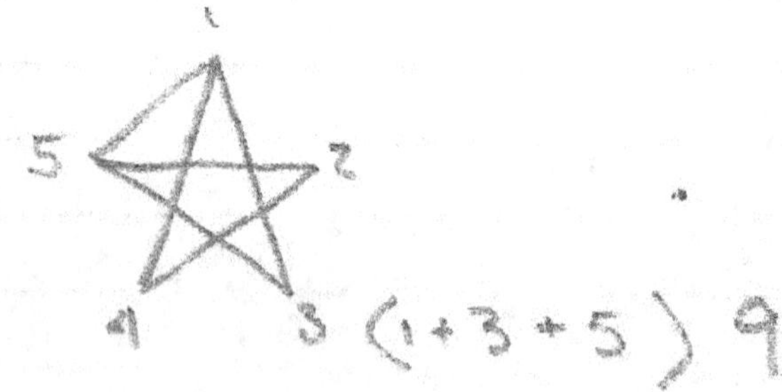

1
5   2
4   3
(1 + 3 + 5) 9

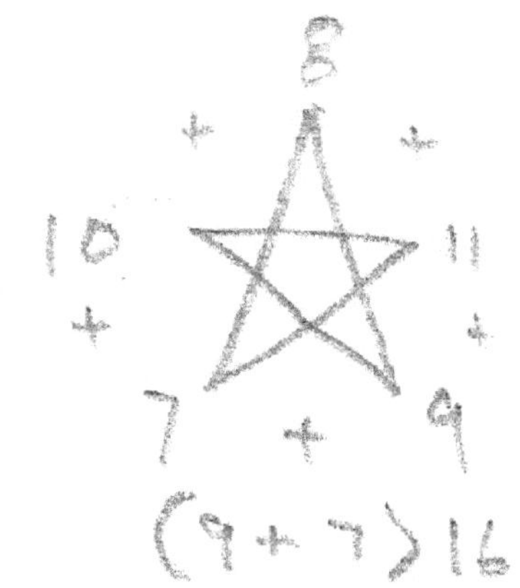

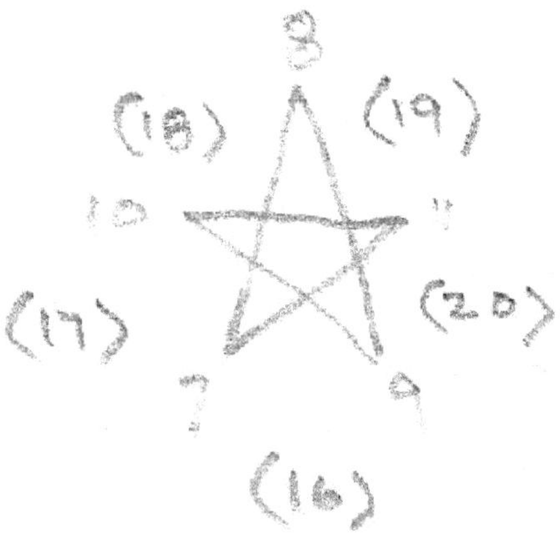

$(4+1)\,5$ $(2+3)\,5$

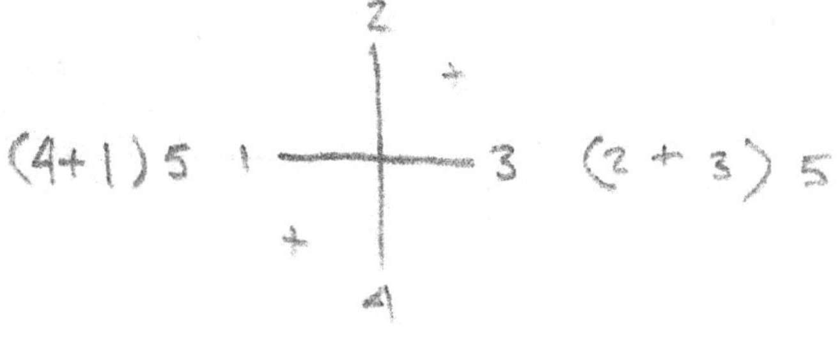

$(6+1)\,7$ $4\cdot(3+4)$

5

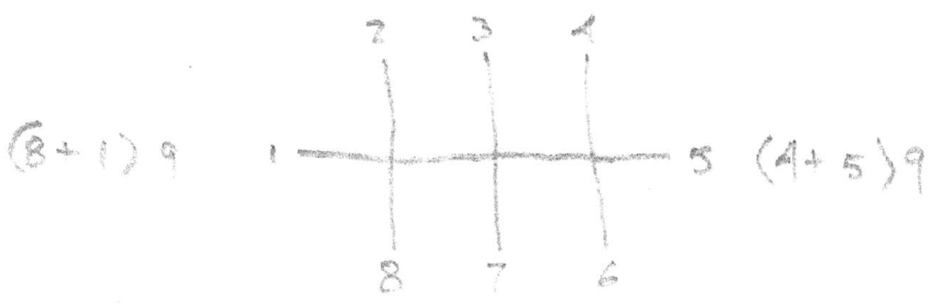

(8+1) 9   1 ————————— 5   (4+5) 9

(4+1) 5

6

(6+1)7

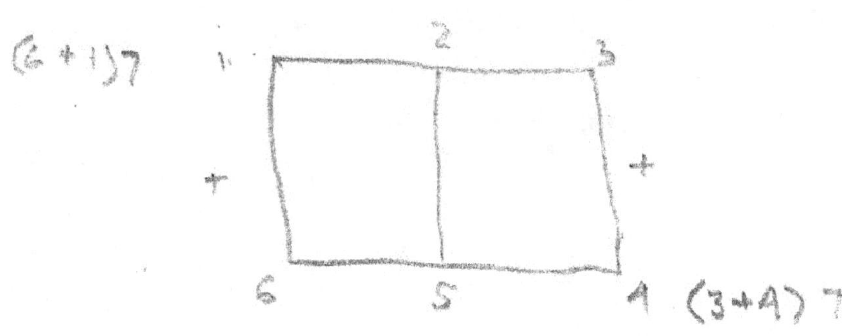

(8+1)9

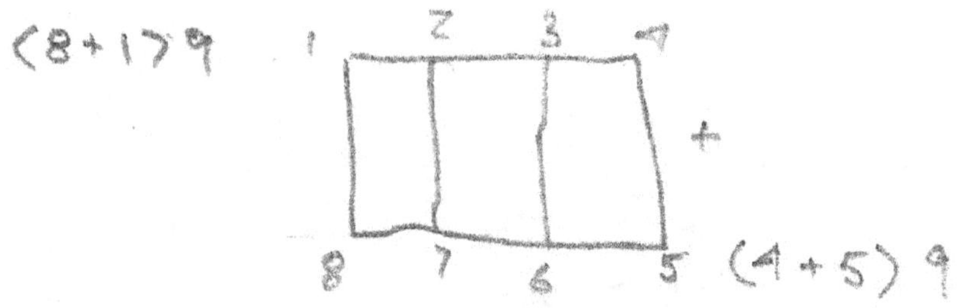

7

1
2
A
3

2
A
3

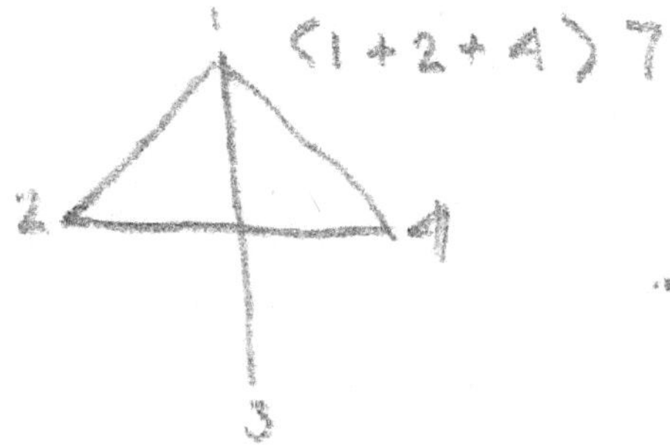
(1 + 2 + 4) 7
2
1
3

8

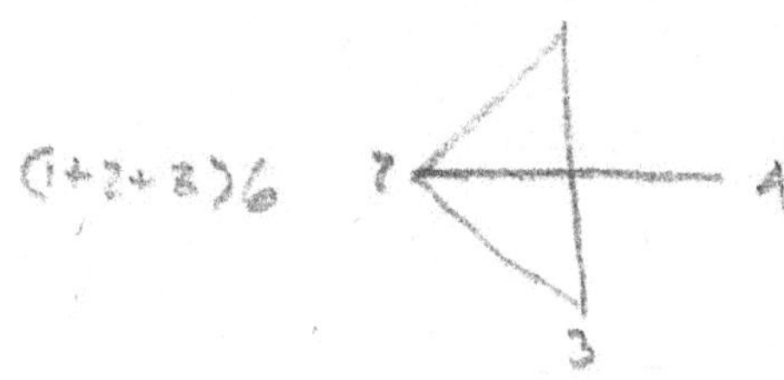

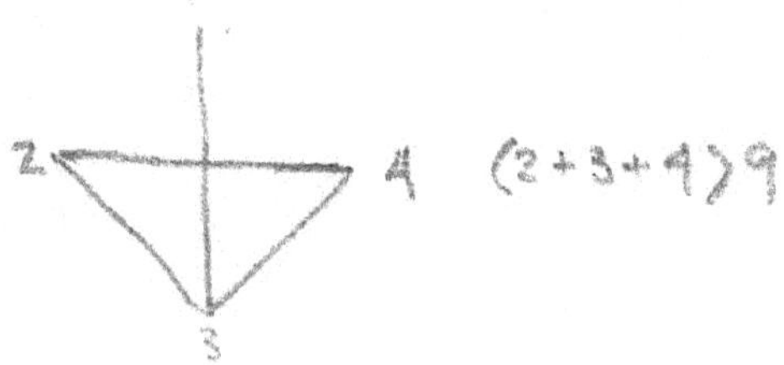

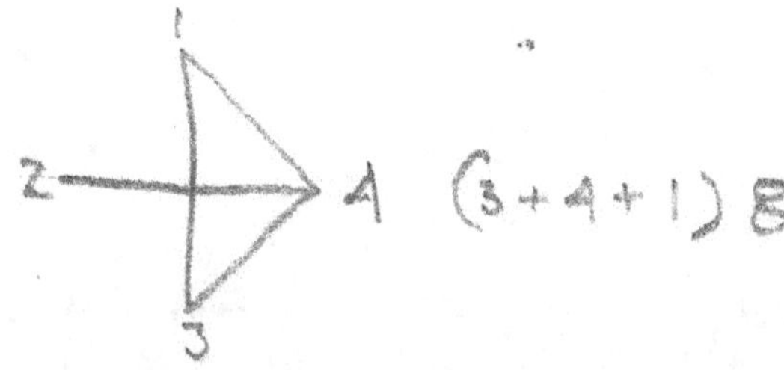

9

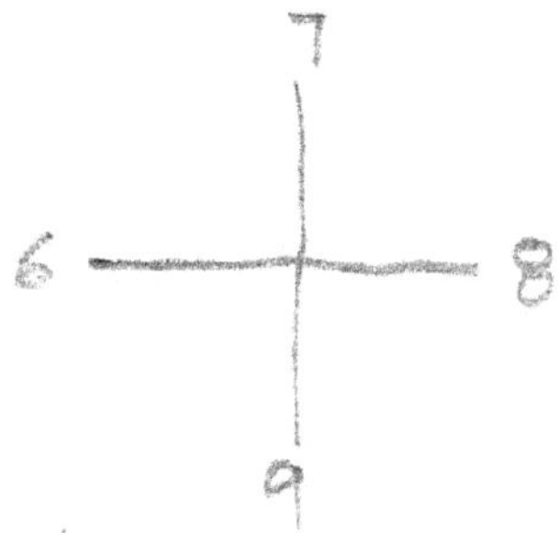

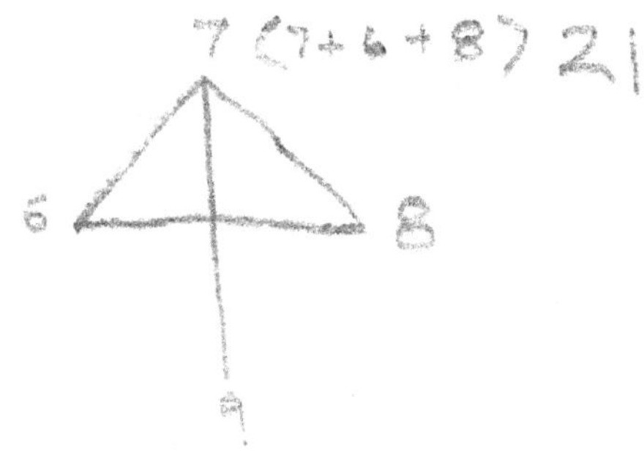

(7 + 6 + 9) 22

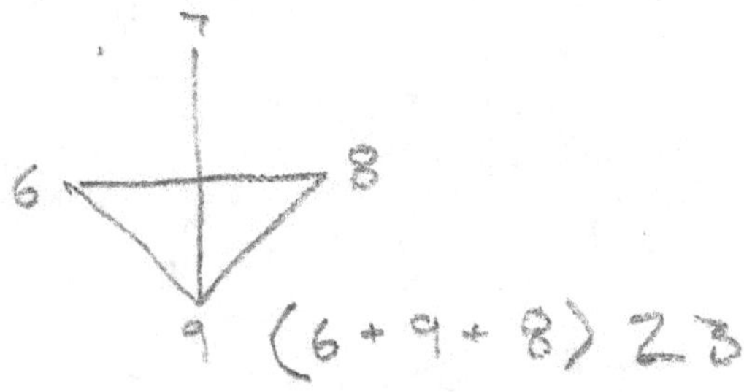

7
6
8
9
(6 + 9 + 8) 23

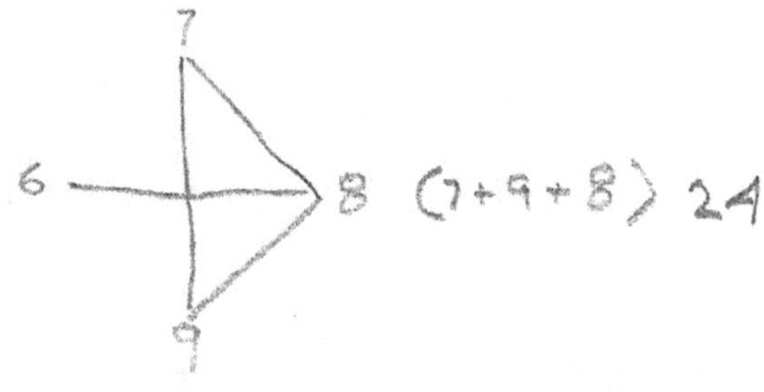

7
6
8
9
(7 + 9 + 8) 24

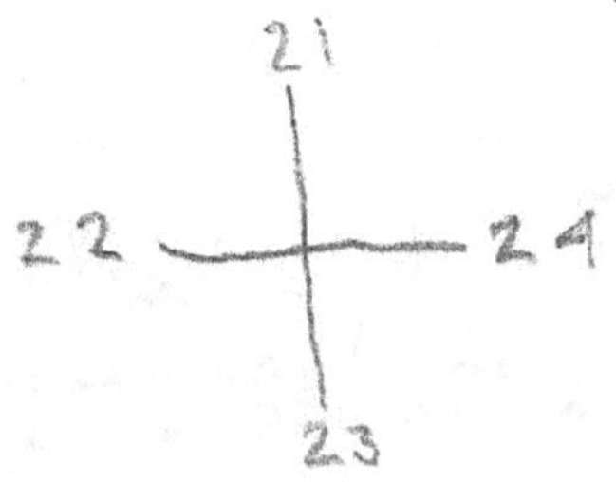

21
22
24
23

www.ingramcontent.com/pod-product-compliance
Lightning Source LLC
Chambersburg PA
CBHW072146150726
48002CB00004B/1647